HOW SEEDS TRAVEL

HOW SEEDS TRAVEL

by Cynthia Overbeck

Photographs by Shabo Hani

A Lerner Natural Science Book

Lerner Publications Company ▪ Minneapolis

Sylvia A. Johnson, Series Editor

Translation of original text by Chaim Uri

Additional research by Jane Dallinger

The publisher wishes to thank Peter D. Ascher,
Professor of Horticultural Science, University of Minnesota,
for his assistance in the preparation of this book.

The glossary on page 46 gives definitions and pronunciations
of words shown in **bold type** in the text.

This edition of this book is available in two bindings:
Library binding by Lerner Publications Company
Soft cover by First Avenue Editions
241 First Avenue North
Minneapolis, Minnesota 55401

LIBRARY OF CONGRESS CATALOGING IN PUBLICATION DATA

Overbeck, Cynthia.
  How seeds travel.

    (A Lerner natural science book)
    Adaptation of: Tane no yukue/Shabo Hani.
    Includes index.
    Summary: Describes how seeds are moved from place
  to place by wind, water, and animals, and how they
  function in plant reproduction.
    1. Seeds—Dispersal—Juvenile literature. [1. Seeds—
  Dispersal. 2. Plants—Reproduction] I. Hani, Shabō,
  ill. II. Hani, Shabō. Tane no yukue. III. Title.
  IV. Series.
  QK929.O93   582'.0467   81-17217
  ISBN 0-8225-1474-5 (lib. bdg.)   AACR2
  ISBN 0-8225-9569-9 (pbk.)

This edition first published 1982 by Lerner Publications Company.
Text copyright © 1982 by Lerner Publications Company.
Photographs copyright © 1978 by Shabo Hani.
Adapted from WHERE THE SEEDS GO copyright © 1978 by Shabo Hani.
English translation rights arranged with Akane Shobo Company, Ltd.,
through Japan Foreign-Rights Centre.

All rights to this edition reserved by Lerner Publications Company.
International copyright secured. Manufactured in the United States of America.

International Standard Book Number: 0-8225-1474-5
International Standard Book Number: 0-8225-9569-9 (pbk.)
Library of Congress Catalog Card Number: 81-17217

7 8 9 10 – P/JP – 98 97 96 95

A yellow dandelion appears in the middle of a smooth green lawn. A wildflower pushes its way up through a crack in a city sidewalk. A tiny maple tree sprouts on a rocky hillside. Why are these plants growing in such places? How did they come to sprout and take root so far away from other plants of their kind?

The dandelion, wildflower, and maple tree have been brought to the places where they are growing by traveling seeds. In this book, you will learn about the many fascinating ways in which plant seeds move from place to place.

Left: In the fall, the wind scatters the seeds of the knotweed plant.

Opposite: In spring, the knotweed seeds that have been protected under the dry grass all winter begin to sprout.

Seeds are the beginning of life for most kinds of plants. A seed contains all the parts necessary to produce a new plant. But in order for a new plant to grow from a seed, certain basic needs must be met. The seed must be in a place where it has good soil and enough water and sunlight to grow.

Sometimes people play a role in helping seeds to grow into plants. They put seeds into the rich earth of gardens or fields, water them, and give them fertilizer and other plant foods. But plants in their natural state do not need human help in order to reproduce themselves. They have many ways of making sure that their seeds get to just the right places so that they will be able to grow.

Some seeds travel by sailing on the wind or floating on water. Others hitchhike on the fur of animals. All traveling seeds have special features and structures that help them to take advantage of the free rides that nature offers.

6

THE PARTS OF A SEED

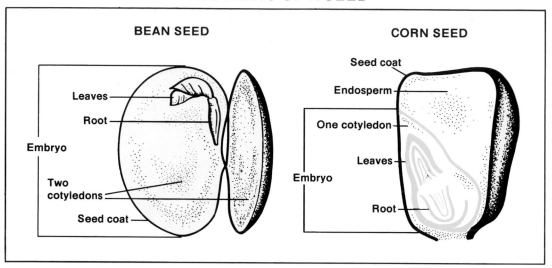

BEAN SEED

Leaves

Root

Embryo

Two cotyledons

Seed coat

CORN SEED

Seed coat

Endosperm

One cotyledon

Leaves

Embryo

Root

Here are seeds from two different kinds of plants. The bean seed is from a *dicot* plant; it has two special structures called *cotyledons* (kaht-'l-EED-unz), which are part of the embryo. The rest of the embryo is made up of the root and the first leaves of the new plant. In a bean seed that is fully developed, food is stored in the two cotyledons. The corn seed has only one cotyledon; it is a *monocot*. Monocot plants store food in a part of the seed called the *endosperm*.

What are plant seeds like, and where do they come from?

All seeds have three main parts: the **seed coat**, the **embryo**, and the **food storage tissue**. The seed coat is an outer skin that protects the tiny embryo inside. The embryo contains all the parts that are needed to make a new plant. The rest of the seed is made up of stored food to feed the embryo.

8

Some flowers are as large and colorful as the sunflower (right), while many others are as small and delicate as the flower of the maple tree (left).

Many of the plants that we see around us grow from seeds that have been produced within flowers. More than half the world's plants bear flowers and are known as **flowering** plants. Garden plants like roses, tulips, and daisies have the bright showy kinds of flowers that are familiar to most people. But many other kinds of plants have flowers, too. Common garden vegetables like the bean and squash plants produce flowers. So do trees like the maple and the cottonwood. Grasses are also flowering plants. Wheat, bamboo, and even common grasses in people's yards have flowers. The flowers on these plants are often simpler, and sometimes much tinier, than garden flowers. They may not have colorful petals or large, bright centers. But all flowers have the basic parts needed to make seeds.

THE PARTS OF A FLOWER

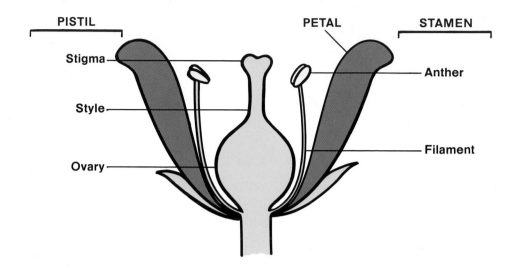

Flowering plants have female and male parts that work together to produce seeds. The seeds grow inside the female part of a flower, which is called the **pistil**. The base of a pistil is a hollow structure called the **ovary**. Inside the ovary are one or more egg cells that can develop into seeds. Usually a **style**, or small stalk, grows up out of the ovary. At the top of the style is the **stigma**, which has a sticky surface.

The egg cells in the ovary can grow into seeds only if they are united with male sperm cells. The part of the flower that produces sperm cells is the **stamen**. Most stamens are made up of a slender stalk called the **filament**, which has an enlarged tip called the **anther**. It is the anther that produces **pollen**, the powdery substance that contains sperm cells.

In order for seeds to be produced, pollen from the anther must reach the stigma at the tip of the pistil. This is called **pollination**. Some plants are **self-pollinating**, which means that pollen is transferred from the anther of one flower to the stigma of the same flower or of another flower growing on the same plant. Plants that are **cross-pollinating** depend on the help of wind or insects to transfer pollen from one plant to another.

After pollination has taken place, pollen grains on the stigma send tiny tubes down the style into the ovary. Sperm cells pass down these tubes and unite with, or **fertilize**, the egg cells in the ovary. After fertilization, the egg cells begin to grow into seeds.

FERTILIZATION IN A FLOWERING PLANT

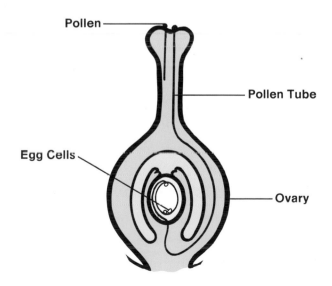

Pollen

Pollen Tube

Egg Cells

Ovary

Inside the brightly colored fruits of the snake gourd plant are seeds that can grow into new plants.

After the egg cell or cells are fertilized, they begin to grow inside the ovary. Usually the petals, style, and stigma dry up and wither. Now much of the plant's energy goes into feeding the growing seeds.

As the seeds develop, the ovary around them grows too. The ovary ripens into a **fruit**. The fruit holds and protects the seed or seeds within.

Many fruits are fleshy and juicy when they ripen. These are the familiar fruits like apples, peaches, or oranges, which people enjoy eating. But other kinds of fruits are dry when they are ripe. Some of these are not what most people usually think of as fruit. They include pea pods, the kernels of corn or wheat, and some nuts like the acorn or the chestnut. A fruit is any part of a flowering plant that holds the seeds.

Right: The bright flowers of the moss rose. *Below:* This enlarged picture shows the round fruits of the moss rose split open. Inside each fruit are many tiny seeds.

The pictures on these two pages show the development of the cosmos plant from flower to fruit.

Like many plants, the cosmos produces flowers during the warm summer months. Cosmos flowers have white, pink, or reddish petals surrounding a yellow center (left). Because the cosmos belongs to a group known as **composite** flowers, it has a different structure than flowers like roses or lilies. The yellow center of each flower head is actually made up of many tiny flowers, which you can see in the cut-away picture on the top right. Each little flower has an ovary containing one egg cell.

After the egg cells in the ovaries have been fertilized, the other parts of the flower head dry up (bottom right). The ovaries and the seeds inside them begin to grow. Before long, the ovaries have developed into fruits.

By autumn, the dry fuits of the cosmos plants are ripe. The seeds inside them are fully developed and ready to begin life on their own.

Once the seeds and the fruits around them are ripe, it is time for the seeds to be released from the parent plant. Then they may sprout and grow into new plants. The **dissemination**, or scattering, of seeds is one of the most important steps in the life cycle of a plant.

Seeds need space if they are to sprout and grow successfully. If all the seeds of a tree, for example, fell right under the parent tree, they would be too crowded. Most of those that did sprout would probably never become full-grown trees able to make their own seeds. This is because they would not receive enough sunlight and food. The tall parent tree would block out much of the sun and use up most of the water and food in the soil. So it is important that at least some of a plant's seeds travel to a place where they will have plenty of room to grow.

The parts of plants that travel are called the **disseminules**. In the plants shown in this book, most of the disseminules are the seeds and fruits themselves. Sometimes the surrounding parts of the flower form special structures that are also part of the disseminule.

Disseminules have several basic ways of traveling. One of the most common ways is by riding the wind. Disseminules that are designed to ride the wind must have certain characteristics. Most important, they must be lightweight. They must also have a lot of surface area compared to their weight. The wind pushes against the surface of such disseminules in much the same way that it pushes against the sail of a sailboat.

Left: A Japanese maple tree in flower. *Above*: An individual maple flower. The green part of the ovary will ripen into the little wings of the fruit.

Many disseminules have "wings" to catch the wind. The fruit of the maple tree is an example of a winged disseminule.

In spring, little flowers bloom on the maple tree. When the two egg cells in the pistil of each flower are fertilized, they begin to grow into seeds. All through the summer, both the seeds and the ovary around them grow. The ovary and its surrounding parts ripen into a dry fruit that branches into two light, papery wings. Each wing has a seed stored at one end. This fruit is the disseminule of the maple tree.

In autumn the maple disseminules begin to fall from the tree. Often the fruit breaks apart so that each half has only one wing and one seed. The breeze catches the wings, and they twirl round and round in the air. This twirling action helps to slow down their fall. The longer they are in the air, the better chance they have of catching a breeze that will carry them away.

A cluster of ripe Japanese maple fruits

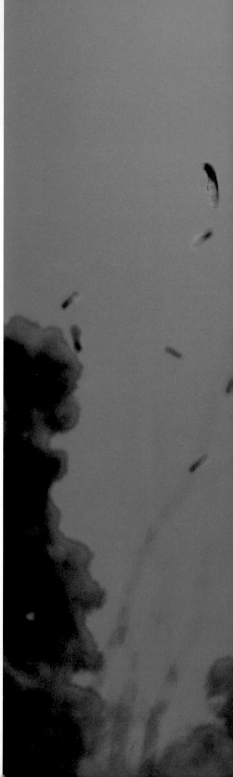

Many seeds will still fall close to the parent tree. But others may travel further. There are records of seeds that have traveled up to 32 miles (50 kilometers) in very strong winds. More often, they are carried a few hundred feet (about 90 meters) before they fall to the ground. Most winged disseminules grow on tall trees or on climbing vines. Growing high up, they have a better chance of catching the wind than if they were nearer the ground.

The maple fruit will lie all winter long on the ground where it landed. When spring comes, the seed may sprout into a little maple tree seedling. If the seedling gets enough light, space, water, and food, it will someday grow into a mature tree and produce its own seeds.

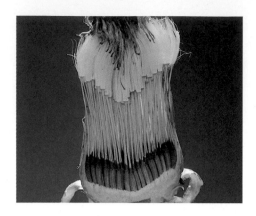

Left: Dandelion flowers in three stages: in bloom; covered with white disseminules; with all disseminules gone. *Above*: Seeds clustered together on a dandelion head.

Winged disseminules like that of the maple usually travel fairly short distances. But other kinds of wind disseminules often travel much farther. These are the disseminules with hairs. Such downy disseminules are so light that the slightest breeze can pick them up and carry them far away.

The dandelion is a familiar flower that has this kind of disseminule. Like the cosmos flower, the dandelion is a composite flower. The flower head is really a cluster of many tiny yellow flowers. Each flower contains its own ovary, with an egg cell that can grow into a seed. When these cells have been fertilized and the other flower parts have dried up, there are many tiny seeds on the dandelion head. Each seed is inside a dry little fruit that has its own downy "parachute" growing upon a thread-like stalk. The parachute is made up of many fine white hairs.

Anyone who has ever blown on a dandelion head covered with these tiny parachutes knows that these disseminules fly off quite easily. When the seeds and fruits are ripe, the dandelion stem stands up straight and tall so that the wind can catch the disseminules. It carries them a good distance — sometimes many miles — before they land.

Right: A single clematis disseminule. *Left*: A clematis vine covered with disseminules.

Disseminules with hairs come in many different shapes. The disseminule of the clematis (klih-MAT-us), a flowering vine, looks like a kind of tiny kite. The fruit with the seed inside is attached to a "tail" that has developed from the style of the clematis flower. The tail has many feathery hairs. Wind can pick up the clematis disseminule and carry it many miles before it lands on the ground.

Cattails also produce downy disseminules. In early fall, the tall stems of the cattail seem to be topped by long, brownish cylinders. These are really clusters of tiny, tightly packed flowers in which seeds are growing. When the seeds and surrounding fruits are ripe, they have downy hairs. Thousands of these cottony disseminules fly off the cattails like a snowdrift in the wind.

24

Cattail disseminules fly away with the wind.

The pictures on these two pages show some of the ways in which traveling seeds may be lost while on their journeys.

Plants like the cattail, dandelion, and clematis produce many, many seeds. They must produce so many because a large number will be lost. Only a few will reach a suitable place, sprout, and grow to maturity.

Seeds may meet with all kinds of accidents on their journeys. Some will be collected and eaten by insects and animals. In the pictures on the left, black ants are dragging the disseminules of a dandelion into their underground storehouse. These seeds will never sprout. Other seeds may never reach the ground. They may be caught in a spider web, for example, and die there (above left). Some seeds will land in dry areas where there is not enough water for them to sprout. Or they will land in streams or lakes and become too wet (above right). Then they may rot before they can sprout.

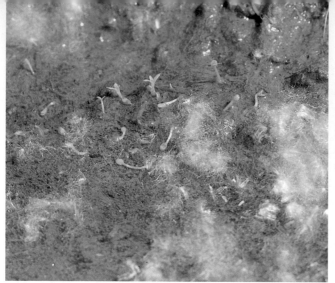

In some cases, water can actually be helpful in spreading seeds. The downy disseminules of the pussy willow, for example, use both wind and water for distribution.

Many pussy willow bushes grow at the edges of ponds or streams. The tiny fruits of the pussy willow are clustered together in little bunches on the bare branches, much like those of the cattail. When they are ripe, they have fine, silky white hairs. The wind blows these puff-like disseminules off the branches and onto the surface of the water. There they float. If the water currents wash them to a good spot on the shore, and if they soak up just the right amount of water during their journey, they will sprout, take root, and grow.

28

Far left: The flowers and disseminules of the sea chamomile. *Left:* Chamomile plants at high tide. *Below:* The disseminules of the chamomile floating on the water.

Left: A sea lavender plant at low tide. *Right*: The same plant underwater at high tide.

Another plant that spreads by using both wind and water is the sea chamomile (KAM-eh-mile) plant. The type of chamomile shown in the pictures on the left grows in marshy areas that are close to the ocean. When the ocean tide comes in, the marsh is flooded. The wind scatters the downy disseminules of the chamomile over the water's surface. When the tide goes out, the movement of the water in the marsh carries away the disseminules with their seeds and distributes them in various places.

The sea lavender plant also depends on water to distribute its seeds. Like the chamomile, the plant grows in salt marshes near the ocean. When the tide rises, this whole plant is underwater. The disseminules may be detached from the plant and carried away by the water as the tide goes out. The seeds will sprout in the water. Then, when the seedlings land in the soil, their roots will dig in.

Many downy disseminules can float on the water because they are so light. Some winged disseminules can also float because they are lighter than the water. These are all fairly small disseminules. But some larger ones can float, too. This is possible if they have air spaces inside them that make them lighter than the water.

One such disseminule is the coconut. The coconut is the fruit of the coconut palm. This tree grows in warm tropical climates. The hard, woody shell of the coconut fruit is as big as a football. Inside is a very large seed. But between the outer covering and the seed are spongy fibers that contain a lot of air space. This air space makes the coconut light enough to float.

Coconuts have sometimes been carried for miles on ocean currents. Often the seed inside becomes soaked with sea water and dies. But if it is washed up on a warm beach before this happens, it may sprout. Coconuts first grew only in Central America. But by traveling on ocean currents, the seeds have reached many of the world's tropical islands and coastal areas. Now groves of coconut palms grow where there were none before.

Here are a few plants that produce barbed, pointed, or spiny disseminules: 1. spanish needles; 2. burdock; 3. cocklebur; 4. herb bennet

Disseminules that are light and of a certain shape or design can be carried by currents of air or water. But there are other ways in which seeds are distributed. Animals do a lot to carry and spread seeds around.

Some disseminules have special structures such as hooks, barbs, spines, and points. These help the disseminules to hitchhike. They catch on the fur of an animal that brushes past them or on a bird's feathers. They may also stick to people's sweaters or pants legs as they go walking in fields or

Left: Wedge-shaped tickseed fruits grow in a row with their hooks projecting at the ends. *Right*: Dry tickseed fruits attached to the wool of a sweater.

forests. Usually people and animals do not notice that they have picked up the disseminules. They may travel a long way before the disseminules fall off or are picked off and thrown onto the ground.

The tickseed fruit has a tiny hook on its end. When tickseed fruits are ripe, they can easily catch on the wool of a sweater or on the hair of a passing dog. Tiny hairs on the outside of the fruit help it to cling, too. Tickseed fruits can be carried many miles before they fall to the ground.

Some disseminules have sticky substances on them that help them to cling to passing animals. The little droplets that look like water on the plant in the picture above are really a sticky liquid that helps the spiny little disseminules to catch a ride.

Without knowing it, animals carry many disseminules on their fur or feathers or in mud on their feet or claws. They also help to distribute seed by deliberately planting the disseminules. Squirrels and blue jays, for example, often collect nuts and bury them to eat later. A large number of these nuts

get lost or forgotten. In the spring, they may sprout where they were buried the previous fall and take root.

Even when birds and animals actually eat fruits containing seeds, they may be helping to get the seeds distributed and planted. The seeds inside fleshy fruits like apples or cherries cannot sprout until the fruits are removed. This is where birds and animals can be helpful.

Often a bird will eat the juicy pulp of a cherry, for example, and drop the seed on the ground. The seed may then sprout and grow. Even when the animal swallows both the pulp and the seed, the seed may still be distributed. It often happens that the seed itself passes whole through the animal's digestive tract. Then it falls to the ground in the bird's or animal's droppings.

One fruit, that of the mistletoe, is distributed by birds in a very unusual way. Mistletoe plants grow on the trunks and branches of trees such as the oak and the hackberry. Mistletoe is a **parasite**; that is, it gets all its nourishment from the tree on which it grows. Its roots grow right into the tree's trunk or branches and draw out the food the mistletoe needs. A mistletoe seed can take root only if it lands on a tree. If it lands on the ground, it will die.

Birds help to make sure that the mistletoe seeds will land in the right place. Some birds like to eat the little round fruits of the mistletoe (above). But the seeds inside the fruits are covered with a substance that makes them stick to the birds' beaks as they eat the fruit. To get the seeds off, the birds wipe their beaks against the bark. Often the seeds stay there and take root.

Sometimes a mistletoe seed is swallowed by the bird along with the fruit. But the seed can go through the digestive tract and not lose its stickiness. Then it may come out in the bird's droppings and stick to the tree branches as it falls.

Above left: Mistletoe growing on a Chinese hackberry tree. *Above right*: A sticky mistletoe seed clings to the branch of a tree. *Below*: Mistletoe sprouts from a tree trunk.

In all of the plants that have been shown so far in this book, the disseminules need some help, either from the wind, water currents, or animals, to get from one place to another. But some plants have special ways of scattering seeds by themselves, without outside help. The wood sorrel is one such plant.

After the yellow flowers have bloomed and the egg cells in the pistils have been fertilized, the fruit that develops is a long, podlike structure. Inside the fruit, the seeds of the wood sorrel are lined up in much the same way that peas are arranged in a pea pod. As the inner walls of the fruit ripen, they begin to dry up. At a certain point they bend backward suddenly, forcing the seeds to fly out. These seeds do not usually travel a great distance. But they do scatter widely enough in the area nearby to find room to sprout. Since they have a rough outer coat, they may stick to an animal's fur or feet and be carried further.

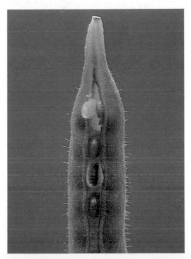

Right: Seeds of the wood sorrel lined up inside the fruit. *Below*: The seeds are suddenly tossed out of the wood sorrel fruit.

Left: The bright flowers of the touch-me-not. *Above*: Many green fruits ripen on the touch-me-not plant.

Another plant that scatters its seeds by tossing them out is the touch-me-not, sometimes called the jewelweed. When its bright red flowers are finished blooming, pod-like fruits about an inch (2.5 centimeters) long develop. Each fruit contains about 20 tiny seeds. When the fruit is ripe, it separates into five sections. These sections coil up suddenly and send the seeds shooting out.

Touch-me-not is a good name for this plant. Once the fruits and seeds are ripe, the fruit will burst open at the slightest touch. The seeds may fly as far as seven feet (about two meters).

The fruits of the touch-me-not split open when they are ripe and toss the dark seeds out with sudden force.

Another plant that splits suddenly and shoots out its seeds is the wild geranium. The green fruits seen in the picture below are still closed around the seeds. The dried-up fruits with the curly arms have already split open like those of the touch-me-not and shot out their seeds.

The mountain wisteria shown on the opposite page also shoots out its seeds. After the lavender flowers have bloomed and the fruit pods are dry, the pods twist and split, giving off a loud cracking sound. The seeds may be shot up to 17 feet (5 meters) away.

We have seen that seeds have many different ways of traveling. They use wings and parachutes of down to fly on the wind. They have hooks and barbs that cling to passing animals. They float on water and shoot into the air.

Because seeds are always traveling, new plants spring up everywhere. As long as seeds move from place to place, weeds, wildflowers, and new trees will continue to sprout and grow, keeping the earth green and alive.

GLOSSARY

anther—the part of a flower's stamen that contains pollen

composite flower—a flower whose head is made up of many small flowers

cross-pollination—the transfer of pollen from the flowers of one plant to the flowers of another plant

dissemination (dis-em-eh-NAY-shun)—the spreading or scattering of seeds

disseminule (dis-EM-eh-nyool)—the part of a plant that travels from one place to another

embryo (EM-bree-oh)—the part of a seed that develops into a new plant

fertilization—the uniting of a male sperm cell and a female egg cell to produce a seed

filament—the stalk of a stamen

flowering plant—a plant that produces flowers, fruits, and seeds

food storage tissue—the part of a seed containing stored food to feed the embryo

fruit—a ripened ovary; the part of a plant that holds seeds

ovary—the part of the pistil in which a seed grows

parasite (PAIR-uh-site)—a plant that takes its food from another living plant

pistil—the female, seed-producing part of a flower

pollen—the powdery substance produced by flowers that contains male sperm cells

pollination—the process by which pollen comes together with the stigma of a flower

seed coat—the outer covering of a seed

self-pollination—the depositing of pollen from a flower onto its own stigma or onto the stigma of another flower on the same plant

stamen (STAY-mehn)—the male, pollen-producing part of a flower

stigma—the part of the pistil that receives pollen

style (STILE)—the narrow part of the pistil that supports the stigma

INDEX